CW00525267

Law of Attraction Simplified

Simplified

A Beginner's Guide to Manifesting Abundance, Love, and Success through Easy Techniques

Amelia Earhart

Contents

Introduction

The Law of Attraction has existed for centuries, but its recent popularity can be attributed to its connection with personal development and spirituality. I have personally experienced its power and believe that it can help anyone manifest their desires and goals into reality.

While the Law of Attraction is often associated with manifesting abundance, love, and success, it's much more than that. This book aims to simplify and demystify the Law of Attraction, providing practical tips and techniques to help you apply them to your life.

At its core, the Law is based on the principle that everything is energy, and our thoughts and beliefs have a powerful effect on the world around us. Our thoughts and beliefs are like magnets that attract similar energy vibrations. By focusing on positive thoughts and beliefs, we can attract positive experiences into our lives. Conversely, by focusing on negative thoughts and beliefs, we can attract negative experiences.

The Law of Attraction works in a simple and straightforward way. When I focus my thoughts and beliefs on what I want to manifest, I send out a powerful vibration to the universe. This vibration is like a signal that tells the universe what I want. The universe then responds by sending back similar energy vibrations that match my thoughts and beliefs. In other words, I attract what I focus on.

Throughout this book, I will explore how the Law of Attraction works, why it's important, and how you can use it to manifest abundance, love, and success. I will start by covering the basics of the Law, including its principles and mechanisms. I'll also discuss the scientific basis behind the Law of Attraction and how it relates to quantum physics.

From there, I will delve into various techniques and practices that can help you harness the power of the Law of Attraction and integrate it into your daily routine. These techniques include visualization, affirmations, gratitude, meditation, and more. I'll also discuss the importance of taking inspired action and how to overcome common blocks and obstacles that may be preventing you from manifesting your desires.

My goal is to provide you with insights and techniques that inspire and empower you to create a life full of abundance, love, and success. This book is for anyone who wants to unlock their full potential and achieve their goals and desires. Whether you're new to the Law of Attraction or have been practicing it for years, this book will offer a comprehensive guide to manifesting your dreams.

I hope that the insights and techniques shared in this book will help you unlock your full potential and create the life you truly desire.

So, let's dive into the principles of the Law of Attraction and learn how to manifest your dreams into reality!

You can also out my published book titled "*369 Manifesting Journal: Your Manifestation Workbook Guided by the Law of Attraction*" for additional manifestation practices. This journal provides a step-by-step guide to help you manifest your goals and desires using the Law of Attraction.

Understanding the Law of Attraction

The Law of Attraction is not just some trendy concept, but it has been around for centuries. For example, the ancient Egyptians and Greeks recognized the power of visualization and positive thinking, and many spiritual traditions incorporate teachings on the Law of Attraction. Even famous personalities like Mahatma Gandhi and Martin Luther King Jr. understood the power of their thoughts in shaping their lives and the world around them.

Gandhi once said,

"Your beliefs become your thoughts, your thoughts become your words, your words become your actions, your actions become your habits, your habits become your values, and your values become your destiny."

In recent times, books such as "The Secret" and "Think and Grow Rich," along with the growth of personal development and self-help movements, have made the Law of Attraction more popular. But the idea is not a fad or a trend, it's a fundamental principle of the universe that can be applied to any aspect of your life, whether it's love, career, health, or personal growth.

So, how does the Law of Attraction work? It all begins with your thoughts and beliefs. Your thoughts act as magnets, attracting experiences and circumstances that match their vibrational frequency. Every thought you have emits a specific vibrational frequency that attracts similar frequencies.

For instance, suppose you are constantly thinking about how much you hate your job and how unhappy you are. In that case, you are likely to attract more experiences that make you feel unhappy and unfulfilled. This is because your negative thoughts emit negative energy vibrations that attract similar negative vibrations in the universe. On the other hand, if you focus on positive thoughts and beliefs, such as how much you enjoy your job and how grateful you are for your blessings, you are more likely to attract positive experiences into your life. Your positive thoughts emit positive energy vibrations that attract similar positive vibrations in the universe.

The Law of Attraction also operates through the power of visualization. When you visualize yourself living your desired reality, you send a powerful message to the universe that you are ready to receive it. Visualization is not just daydreaming or wishful thinking; it's creating a clear mental picture of what you want and holding that vision in your mind. It's like creating a blueprint for your life and taking steps towards it.

Let me share a personal story about the power of visualization. When I was in college, I wanted to travel the world and experience different cultures. But at the time, I didn't have the means or resources to make it happen. So, I started visualizing myself traveling to different countries, experiencing new things, and meeting interesting people. I even created a vision board with pictures of places I wanted to visit and things I wanted to do. Over time, opportunities started to arise, and I was able to travel to different countries and experience the things I had visualized. It was like the universe was aligning with my thoughts and making things happen.

However, the Law of Attraction is not just about positive thinking and visualization. It also requires action. You cannot simply wait for your dreams to come true; you must take inspired action toward your goals. This entails taking steps to align your actions with your desires, such as networking, learning new skills, or taking risks.

For example, let's say you want to start a business. You can't just sit around and hope that customers will magically appear. You need to take action, such as researching your market, developing a business plan, and networking with potential clients. When you take inspired action towards your goals, you create momentum and attract even more positive experiences into your life.

In other words, the Law of Attraction is a combination of positive thinking, visualization, and action. By aligning your thoughts, beliefs, and actions with your goals, you can manifest abundance, love, and success in all areas of your life.

One thing to keep in mind is that the Law of Attraction isn't a magic wand that will instantly grant all your wishes. It requires time, effort, and consistency. Sometimes, you may encounter setbacks or challenges that test your resolve. But if you stay focused on your goals, maintain a positive attitude, and take inspired action, you can overcome any obstacle and achieve your desires.

By focusing on positive thoughts and beliefs, visualizing your desired reality, and taking inspired action towards your goals, you can manifest abundance, love, and success in all areas of your life. With time, effort, and consistency, anything is possible. Remember, as Mahatma Gandhi once said,

"Be the change you wish to see in the world."

By embodying your desires and living your life as if they have already come true, you can attract them into your reality.

How to Use the Law of Attraction

The Law of Attraction is a powerful tool for manifesting your desires, but it's not always easy to use. Many people encounter common challenges that can prevent them from achieving their goals. In this chapter, we'll explore some of these challenges and provide strategies to overcome them.

One of the most common challenges people face when using the Law of Attraction is resistance. Resistance is the feeling of being blocked or stuck, and it often comes from deep-seated beliefs or fears that hold us back. For example, you might have a belief that you're not worthy of success, or a fear of failure that stops you from taking action towards your goals. These limiting beliefs create negative energy that can block the Law of Attraction from working in your favor.

To overcome resistance, it's important to identify the underlying beliefs or fears that are causing it. Take some time to reflect on your thoughts and feelings, and ask yourself where they're coming from. Once you've identified the source of your resistance, you can work on releasing it through techniques like meditation, journaling, or therapy. These practices help you to reframe your beliefs and develop a more positive mindset, which opens the door for the Law of Attraction to work in your favor.

Another challenge is impatience. When we want something badly, it's natural to want it right away. However, the Law of Attraction works on its own timeline, and it may take time for your desires to manifest. Impatience can create negative energy that actually blocks your desires from coming to fruition. Instead of focusing on the lack of what you want, focus on feeling grateful for what you already have.

To overcome impatience, it's important to trust in the process and have faith that your desires are on their way. Practice patience and stay focused on your intentions, even when you don't see immediate results. Remember that the universe has a plan for you, and everything is happening in divine timing. Trust in the journey and enjoy the process. Don't forget to celebrate small wins along the way.

A third challenge is negative thinking. Negative thoughts and beliefs can create a downward spiral of negative energy that blocks your desires from coming to fruition. To overcome negative thinking, it's important to become aware of your thoughts and challenge them when they're not serving you. Replace negative thoughts with positive affirmations and visualize your desired reality.

Whenever you find yourself thinking negatively, try to reframe your thoughts in a more positive way. For example, if you're feeling anxious about a job interview, instead of thinking "I'm never going to get this job," try thinking "I am qualified and capable, and I trust that the right opportunity will come to me." Positive thinking creates positive energy that attracts more positive experiences into your life.

As Oprah Winfrey once said,

"The biggest adventure you can ever take is to live the life of your dreams."

The Law of Attraction can help you achieve this adventure, but it requires effort and dedication. By overcoming common obstacles like resistance, impatience, and negative thinking, you can unlock the full potential of the Law of Attraction and manifest abundance, love, and success in all areas of your life. Remember to trust in the process, stay focused on your intentions, and have faith that your desires are on their way. With these tools in your arsenal, anything is possible.

Common Mistakes in Using the Law of Attraction

Let's explore some of the common misconceptions and mistakes that people make when trying to apply the Law of Attraction in their lives. By avoiding these common mistakes, you can harness the full power of the Law of Attraction to manifest your desires.

One of the biggest mistakes people make when using the Law of Attraction is focusing on what they don't want. This happens because many people are conditioned to focus on the negative and are constantly worried about things going wrong. However, the Law of

Attraction operates based on the principle of "like attracts like," meaning that you attract what you focus on. If you're constantly thinking about what you don't want, you'll end up attracting more of the same. To avoid this, it's important to shift your focus to what you do want and visualize yourself experiencing it in detail.

Another mistake people make is relying too heavily on manifestation techniques without taking any action. Visualization and affirmations can be powerful tools for manifesting your desires, but they're not enough on their own. You also need to take inspired action towards your goals, such as making plans and taking steps towards achieving your desires. Without action, your manifestations may never come to fruition.

Having a negative mindset is another common mistake people make when using the Law of Attraction. Negative thoughts and beliefs can create energetic blocks that prevent you from manifesting your desires. It's important to cultivate a positive mindset and focus on the things you're grateful for, rather than dwelling on the negative aspects of your life. By doing so, you'll create a positive energetic frequency that will help to attract more positivity into your life.

Another mistake people make is expecting instant results. While the Law of Attraction can work quickly in some cases, it often takes time for your desires to manifest. It's important to be patient and trust that the universe is working behind the scenes to bring you what you desire. Trusting in the process and letting go of the outcome can be challenging, but it's an essential part of using the Law of Attraction effectively.

Finally, it's important to avoid getting too attached to a specific outcome. While it's good to be clear about what you want, being too attached to a specific outcome can create resistance and prevent you from manifesting your desires. Instead, focus on the feeling you want to experience, and trust that the universe will bring you the best possible outcome. By staying open to all possibilities and letting go of the need to control the outcome, you'll create space for the universe to work its magic.

By avoiding these mistakes and staying focused on your desires, you can harness the power of the Law of Attraction to create the life you truly desire. Remember to stay positive, take inspired action, and trust in the universe to bring you what you desire. With practice and perseverance, you'll be able to manifest your desires and live the life you've always dreamed of.

Overcoming Resistance and Blocks

The Law of Attraction can be a powerful tool for manifesting your desires, but sometimes you may encounter resistance that prevents you from achieving your goals. Resistance can take many forms, including limiting beliefs, past traumas, negative self-talk, and fear of failure or success. In this chapter, we will explore some effective ways to overcome resistance and manifest your desires.

Limiting beliefs are beliefs that we hold about ourselves and the world around us that limit our potential for success. These beliefs can be related to our self-worth, abilities, or societal expectations. For example, if you believe that you are not smart enough or deserving enough to achieve success, you may find it difficult to manifest abundance in your life. To overcome limiting beliefs, you need to identify them and work to change them. One way to do this is to practice positive affirmations that counteract your negative beliefs. Another way is to examine your beliefs and challenge them with evidence that contradicts them.

Past traumas can also create resistance by blocking the flow of positive energy and preventing you from manifesting your desires. Traumatic experiences can leave you feeling fearful, anxious, and doubtful of your own power and worthiness. To overcome past traumas, you need to work through them and release the negative emotions associated with them. One effective way to do this is to seek therapy or counseling to process your feelings and gain new perspectives on your experiences.

Negative self-talk is another common form of resistance that can hold you back from achieving your goals. When you constantly criticize yourself or doubt your abilities, you create a negative energy that repels positive opportunities and experiences. To overcome negative self-talk, you need to practice self-compassion and positive self-talk. You can do this by affirming your worth and abilities, focusing on your strengths, and celebrating your successes.

Fear of failure or success is another common form of resistance that can prevent you from manifesting your desires. When you fear failure, you may avoid taking risks or pursuing your goals. When you fear success, you may sabotage your own efforts or hold yourself back from achieving your full potential. To overcome fear of failure or success, you need to acknowledge your fears and work through them. One effective way to do this is to take small steps towards your goals and celebrate your progress along the way.

Resistance is a natural part of the Law of Attraction process, but it doesn't have to hold you back. By identifying the sources of your resistance and working to overcome them, you can manifest your desires more easily and achieve your goals. Remember to practice self-compassion, challenge your limiting beliefs, and take small steps towards your goals. With time and effort, you can overcome resistance and manifest the life you truly desire.

Using the Law of Attraction for Relationships

Let's explore how to apply the Law of Attraction to various types of relationships, including romantic relationships, friendships, and family dynamics.

Firstly, it's important to set clear intentions for the type of relationships we want to attract. This can involve identifying the qualities and values we're looking for in a partner or friend, or the type of dynamic we want to create within our family. By visualizing and affirming our desires, we can send a clear signal to the universe and attract the people and situations that align with our goals.

In romantic relationships, setting boundaries is key to maintaining a healthy dynamic. By communicating our needs and expectations clearly, we can avoid misunderstandings and conflicts. Affirmations can also be useful for improving communication and building trust. For example, you can affirm that you and your partner have open and honest communication, or that you trust each other completely.

In friendships, the Law of Attraction can help us attract like-minded and supportive individuals into our lives. By focusing on positive qualities such as kindness and generosity, we can attract friends who share those same qualities. Affirmations can also be used to strengthen existing friendships and attract new ones. You can affirm that you have meaningful and fulfilling friendships, or that you attract friends who uplift and inspire you.

Lastly, the Law of Attraction can also be applied to family dynamics. By focusing on positive qualities such as love and understanding, we can improve our relationships with family members and create a more harmonious home environment. Affirmations can be used to foster positive interactions and promote forgiveness and healing.

The Law of Attraction can be a valuable tool for improving our relationships with others. By setting clear intentions, using affirmations, and focusing on positive qualities, we can attract loving and supportive relationships into our lives.

Using the Law of Attraction for Health and Well-being

The Law of Attraction is not only about manifesting material possessions and success, but it can also be used to improve our health and well-being. Let's discuss the power of positive thinking and how it can influence our physical and mental health. We will also explore how we can use the Law of Attraction to achieve a healthier, happier life.

The Connection between Thoughts and Health

Our thoughts and emotions have a powerful effect on our physical bodies. Negative emotions such as stress, anger, and anxiety can cause a wide range of health problems, including high blood pressure, heart disease, and depression. On the other hand, positive emotions such as love, joy, and gratitude can boost our immune system, lower our stress levels, and improve our overall health.

The Law of Attraction teaches us that we attract what we focus on, whether positive or negative. Therefore, by focusing on positive thoughts and emotions, we can attract more health and well-being into our lives. This means that if we want to improve our physical health, we need to start by improving our mental and emotional health.

Tips for Using the Law of Attraction for Health and Well-being

1. Focus on positive affirmations: Affirmations are positive statements that we repeat to ourselves to reinforce positive beliefs and attract positive experiences into our lives. Some examples of health-related affirmations include: "I am healthy and strong," "I love my body and take care of it," and "I am grateful for my good health."

2. Visualize yourself in perfect health: Visualization is a powerful tool for manifesting our desires. By visualizing ourselves in perfect health, we can create a mental image of what we want to manifest in our physical reality. Close your eyes and imagine yourself feeling strong, vibrant, and full of energy.

3. Practice gratitude: Gratitude is a powerful emotion that can help us attract more positive experiences into our lives. By focusing on what we are grateful for, we can shift our focus from

negative to positive thoughts and emotions. Take a few minutes each day to reflect on what you are grateful for in your life.

4. Take action towards your health goals: The Law of Attraction is not just about positive thinking, it also requires us to take action towards our goals. Whether it's eating healthier, exercising more, or seeking medical treatment, taking action towards our health goals is an important part of manifesting good health.

By using the Law of Attraction, we can improve our health and well-being in a holistic way. By focusing on positive thoughts and emotions, visualizing our desired outcomes, and taking action towards our health goals, we can manifest the healthy and happy life we desire. Remember that the Law of Attraction is not a magic pill that will instantly cure all our health problems, but it is a powerful tool that can help us achieve a healthier, happier life.

Using the Law of Attraction for Career and Finances

Our career and finances are important aspects of our lives that can greatly impact our overall well-being and happiness. By using the Law of Attraction, we can align our thoughts, beliefs, and actions with our desired career and financial goals. In this chapter, we will explore some practical tips for using the Law of Attraction to improve our professional lives.

1. Set Clear Goals: The first step in using the Law of Attraction for our career and finances is to set clear and specific goals. This includes both short-term and long-

term goals. By having a clear vision of what we want to achieve, we can focus our thoughts and actions towards those goals. It's important to make sure our goals are realistic and achievable.

2. Visualize Success: Visualization is a powerful tool for manifesting our desires. By visualizing ourselves already achieving our career and financial goals, we can align our thoughts and beliefs with that reality. Visualization helps us to create a positive mental image of our desired outcomes, which can help to attract them into our lives.

3. Take Inspired Action: While visualization is important, it's not enough on its own. We also need to take inspired action towards our goals. This means taking steps towards our goals that are in alignment with our thoughts and beliefs. For example, if we want to start a business, we need to take action towards setting up the business and promoting it to potential customers.

4. Release Limiting Beliefs: Our beliefs about money and success can greatly impact our ability to achieve our career and financial goals. If we have limiting beliefs about money, such as "money is the root of all evil," or "I'm not good enough to be wealthy," we will struggle to attract financial abundance into our lives. By releasing these limiting beliefs and replacing them with positive beliefs about money and success, we can open ourselves up to greater opportunities.

5. Practice Gratitude: Gratitude is a powerful way to shift our focus from lack to abundance. By focusing on what we are grateful for in our career and finances, we can attract more of those positive experiences into our lives. For example, if we are grateful for the opportunities we have in our current job, we are more likely to attract more opportunities and promotions in the future.

In summary, by using the Law of Attraction, we can align our thoughts, beliefs, and actions with our desired career and financial goals. By setting clear goals, visualizing success, taking inspired action, releasing limiting beliefs, and practicing gratitude, we can create a positive and abundant career and financial future for ourselves.

Advanced Techniques for Using the Law of Attraction

In the previous chapters, we discussed the basics of the Law of Attraction and how to use it to manifest abundance, love, and success in our lives. In this chapter, we will explore more advanced techniques for using the Law of Attraction that can help us achieve even greater results.

Meditation

One of the most powerful tools for manifesting our desires is meditation. Meditation helps us to quiet our minds, focus our thoughts, and connect with our inner selves. When we are in a meditative state, we can more easily visualize our desires and align ourselves with the energy of abundance.

To start a meditation practice, find a quiet and comfortable space where you won't be disturbed. Sit with your back straight and your eyes closed, and focus on your breath. Allow your thoughts to come and go without attaching to them, and try to stay present in the moment.

Visualization

Visualization is another powerful technique for using the Law of Attraction. By visualizing our desired outcomes, we can more easily attract them into our lives. Visualization is a form of mental rehearsal that helps us to create a vivid picture of what we want to manifest.

To practice visualization, find a quiet space where you won't be disturbed. Close your eyes and picture yourself in a scene that represents your desired outcome. Use all of your senses to make the visualization as vivid as possible, and allow yourself to feel the emotions that come with it.

Energy Work

Energy work is another technique that can help us to align ourselves with the energy of abundance. Energy work involves working with the chakras and the energy centers of the body to remove blockages and improve the flow of energy.

There are many different forms of energy work, including Reiki, acupuncture, and EFT (Emotional Freedom Technique). Each of these techniques can help us to release negative energy and align ourselves with the positive energy of abundance.

Gratitude

Gratitude is a powerful tool for manifesting our desires. When we are in a state of gratitude, we are more open to receiving abundance and blessings in our lives. By focusing on what we are grateful for, we can attract even more things to be grateful for.

To practice gratitude, make a daily practice of writing down things that you are grateful for. This can be anything from a good cup of coffee to a loving relationship or a successful business deal. By focusing on what we are grateful for, we can attract more abundance into our lives.

These advanced techniques can help us to achieve even greater results with the Law of Attraction. By incorporating meditation, visualization, energy work, and gratitude into our daily lives, we can more easily align ourselves with the energy of abundance and manifest our desires.

Maintaining a Law of Attraction Mindset

As we continue to apply the Law of Attraction in our lives, it is essential to maintain a positive and abundant mindset. This means staying focused on our intentions and goals, even when faced with challenges and setbacks. In this chapter, we will explore some practical tips for maintaining a Law of Attraction mindset.

1. Self-Care: The first step in maintaining a Law of Attraction mindset is taking care of ourselves. This includes getting enough rest, eating healthy, and exercising regularly. When our bodies

are healthy, our minds are also more positive and focused.

2. Self-Reflection: Another essential aspect of maintaining a Law of Attraction mindset is self-reflection. This means taking the time to examine our thoughts, emotions, and beliefs. By becoming aware of our inner world, we can identify any limiting beliefs or negative thought patterns that may be blocking our manifestation efforts.

3. Self-Awareness: Self-awareness is closely linked to self-reflection. By being aware of our thoughts and emotions in the present moment, we can better control our reactions and maintain a positive mindset. This means staying mindful and present in each moment and not allowing negative thoughts to take hold.

4. Gratitude: Gratitude is a powerful tool for maintaining a Law of Attraction mindset. By focusing on the good things in our lives and being grateful for them,

we attract more positive experiences. We can practice gratitude by keeping a gratitude journal, expressing gratitude to others, or simply taking time each day to reflect on the good things in our lives.

5. Positive Affirmations: Using positive affirmations is another effective way to maintain a Law of Attraction mindset. By repeating positive statements to ourselves, we can shift our beliefs and focus our thoughts on what we want to manifest. Affirmations can be used throughout the day, such as during meditation or while driving.

6. Consistency: Consistency is key in maintaining a Law of Attraction mindset. It is important to practice these techniques regularly and make them a part of our daily routine. By consistently focusing on our intentions and maintaining a positive mindset, we can manifest our desires more easily.

Maintaining a Law of Attraction mindset is essential for manifesting abundance, love, and success in our lives. By practicing self-care, self-reflection, self-awareness, gratitude, positive affirmations, and consistency, we can maintain a positive and abundant mindset, even in the face of challenges and setbacks.

The Law of Attraction and Abundance Mindset

The Law of Attraction is closely linked to the abundance mindset. The more we focus on abundance, the more abundance we attract into our lives. But what exactly is an abundance mindset? And how can we cultivate it to manifest more of what we desire?

An abundance mindset is the belief that there is always more than enough to go around. It's a mindset of abundance, rather than scarcity. People with an abundance mindset believe that there is an unlimited supply of everything they need, including love, money, and opportunities.

On the other hand, a scarcity mindset is the belief that there is not enough to go around. People with a scarcity mindset often feel like they are lacking in some way, and they believe that there is a limited supply of the things they need. They may feel jealous or envious of others who seem to have more than they do.

Cultivating an abundance mindset is essential if we want to use the Law of Attraction to manifest our desires. When we focus on abundance, we attract more abundance into our lives. We start to see opportunities where we once saw obstacles, and we begin to feel more positive and optimistic about our future.

One way to cultivate an abundance mindset is to practice gratitude. When we are grateful for what we already have, we start to attract more of what we desire. Gratitude helps us focus on the positive aspects of our lives, rather than the negative ones.

Another way to cultivate an abundance mindset is to surround ourselves with positive, supportive people. When we are around people who believe in us and our abilities, we are more likely to believe in ourselves and our potential for success.

It's also important to let go of limiting beliefs and negative self-talk that can keep us stuck in a scarcity mindset. For example, if you believe that you will never have enough money, you are unlikely to attract more money into your life. By changing your thoughts and beliefs to ones that focus on abundance, you can shift your energy and start manifesting what you desire.

Real-life examples of people who have used the Law of Attraction to cultivate an abundance mindset and attract more abundance into their lives are plentiful. For instance, Oprah Winfrey is known for her abundance mindset, and she attributes her success to her ability to focus on abundance rather than scarcity. Similarly, Tony Robbins, one of the world's most successful motivational speakers, teaches people how to cultivate an abundance mindset in order to achieve their goals and dreams.

By cultivating an abundance mindset and using the Law of Attraction to manifest our desires, we can create a life that is filled with abundance, love, and success. It all starts with a simple shift in our thoughts and beliefs, and a commitment to focusing on abundance rather than scarcity.

The Law of Attraction and Gratitude

Let's dive deeper into the powerful concept of gratitude and its connection to the Law of Attraction. As we have discussed earlier, the Law of Attraction works by bringing into our lives what we focus on and put energy into, whether it is positive or negative. And gratitude is a powerful tool for shifting our focus towards positivity and abundance.

Gratitude is the act of being thankful and expressing appreciation for the good things in our lives. It is not only a polite gesture, but a powerful emotion that can transform our thoughts, emotions, and overall outlook on life. When we practice gratitude regularly, we shift our focus towards the positive aspects of our lives, which in turn attracts more positivity and abundance.

Studies have shown that cultivating a daily gratitude practice can have a profound impact on our well-being. It can increase feelings of happiness, reduce stress and anxiety, and even improve physical health. By expressing gratitude, we activate the brain's reward system and release feel-good hormones like dopamine and serotonin.

In the context of the Law of Attraction, gratitude is a powerful tool for manifesting our desires. When we focus on what we are grateful for, we attract more of those things into our lives. For example, if we express gratitude for our job, we are likely to attract more opportunities for career growth and success. If we express gratitude for our relationships, we are likely to attract more fulfilling and loving relationships into our lives.

One of the best ways to cultivate a gratitude practice is to make it a daily habit. This can be done through journaling, meditation, or simply taking a few minutes each day to reflect on the good things in our lives. Some people like to write down three things they are grateful for each day, while others prefer to mentally list them. It doesn't matter what method we choose, as long as we make it a regular practice.

It is important to note that practicing gratitude does not mean denying or ignoring the challenges and difficulties in our lives. Rather, it is about shifting our focus towards the positive aspects and finding reasons to be grateful in the midst of adversity. By doing so, we open ourselves up to more abundance and positivity, and ultimately attract more of what we desire into our lives.

In the next chapter, we will explore how the Law of Attraction is connected to the concept of self-love, and how loving ourselves can help us attract more love and abundance into our lives.

The Law of Attraction and Self-Love

The Law of Attraction is not just about manifesting material objects or success, it is also about creating a fulfilling and happy life. In order to manifest our desires, we need to have a strong foundation of self-love and self-worth. In this chapter, we will explore the importance of self-love and how it is intimately linked with the Law of Attraction.

Self-love is the foundation of all abundance. When we truly love ourselves, we are able to create a life that is aligned with our deepest desires. We are able to manifest our desires with ease, and we are able to attract the right people, circumstances, and opportunities into our lives. However, cultivating self-love can be challenging, especially in a world where we are constantly bombarded with messages that tell us we are not good enough.

One of the first steps in cultivating self-love is to recognize and challenge the negative self-talk that we engage in. Often, we are our own worst critics, and we may say things to ourselves that we would never say to a friend. For example, we might tell ourselves that we are not smart enough, not attractive enough, or not worthy of love. These negative thoughts can hold us back from achieving our goals and living a fulfilling life.

To counteract negative self-talk, it is important to practice self-compassion. This means treating ourselves with kindness and understanding, rather than criticism and judgment. When we make a mistake or fall short of our goals, instead of beating ourselves up, we can practice self-compassion by acknowledging that we are human and that we are doing the best we can in the moment.

Another important aspect of self-love is self-care. This means taking care of our physical, emotional, and spiritual needs. Self-care can include things like eating well, exercising regularly, getting enough sleep, spending time in nature, and engaging in activities that bring us joy and fulfillment.

In addition to self-care, we also need to develop a sense of self-worth. This means recognizing our own value and worthiness, regardless of external circumstances. We are all inherently worthy of love and happiness, and we don't need to prove our worth to anyone else. When we have a strong sense of self-worth, we are able to set healthy boundaries, pursue our goals and dreams, and attract positive relationships into our lives.

In summary, self-love is an essential component of the Law of Attraction. When we love and value ourselves, we are able to create a life that is aligned with our deepest desires. By practicing self-compassion, self-care, and developing a sense of self-worth, we can cultivate a foundation of self-love that will support us in manifesting our desires and living a fulfilling life.

Maintaining a Consistent Practice

The Law of Attraction is not a one-time effort, but a lifelong practice. It is essential to cultivate a consistent practice to manifest our desires successfully. In this chapter, we will explore how to create a daily practice that works for you and how to stay motivated and accountable in your manifestation journey.

Creating a Daily Practice

Creating a daily practice that works for you is essential to maintain consistency in your manifestation journey. Your practice should be something that resonates with you and is sustainable in the long term. Here are some tips for creating a daily practice:

1. Start small: Start with a practice that is easy to do and fits into your daily routine. For example, you can start by setting aside five minutes each day to meditate or visualize your goals.

2. Be consistent: Consistency is key in developing a daily practice. Make it a habit to practice at the same time every day, and gradually increase the amount of time you spend on your practice.

3. Mix it up: It can be helpful to mix up your practice to keep it interesting and engaging. Try different techniques like meditation, visualization, or journaling, and find what works best for you.

Staying Motivated and Accountable

Maintaining motivation and accountability is essential to keep your manifestation journey on track. Here are some tips for staying motivated and accountable:

1. Keep a manifestation journal: Keeping a journal of your manifestation journey can help you stay accountable and track your progress. Write down your goals, intentions, and any successes or setbacks you encounter.

2. Surround yourself with positivity: Surround yourself with people and things that uplift and inspire you. Avoid negative influences that may discourage you from pursuing your goals.

3. Celebrate your successes: Celebrate each success, no matter how small. Acknowledge your progress and give yourself credit for the work you have done.

4. Stay focused on your goals: Stay focused on your goals and remember why you started your manifestation journey in the first place. Use positive affirmations and visualization techniques to keep your mind focused on your desires.

5. Maintaining a consistent practice is crucial to successful manifestation. By creating a daily practice that works for you and staying motivated and accountable, you can achieve

your goals and manifest your desires. Remember, the Law of Attraction is a lifelong journey, and with patience and persistence, you can create the life you desire.

Putting it All Together

So far, you've learned about the Law of Attraction and how it can be used to manifest abundance, love, and success in your life. You've explored various techniques and strategies, and hopefully, you've started to see some positive changes in your life.

Now, it's time to bring it all together. In this chapter, I'll provide you with a step-by-step guide to using the Law of Attraction in all areas of your life. This guide will help you to create a personalized plan for manifesting your desires, and it will provide you with the tools and resources you need to stay focused and motivated.

To begin, take some time to reflect on your goals and desires. What do you want to manifest in your life? Is it a new job or career? A loving relationship? Financial abundance? Improved health and well-being? Whatever it is, write it down and be as specific as possible.

Next, start to visualize yourself already having achieved your goals. See yourself in that new job, or in a happy and loving relationship, or enjoying financial freedom. Use all of your senses to create a vivid mental image, and feel the emotions of joy, gratitude, and excitement that come with achieving your desires.

Once you have a clear vision of what you want to manifest, it's time to take action. This doesn't mean forcing things to happen, but rather taking inspired action towards your goals. This may mean networking to find new job opportunities, joining a dating site to meet new people, or learning about investing to improve your finances. Trust your intuition and take action on the opportunities that come your way.

Along the way, it's important to stay focused on your goals and to maintain a positive mindset. Practice gratitude daily by focusing on the things in your life that you are thankful for. Use affirmations to reinforce positive beliefs and to counter any negative self-talk. And remember to take care of yourself by practicing self-care and self-reflection regularly.

Finally, know that the Law of Attraction is always at work, whether you are aware of it or not. By consciously using these techniques and strategies, you can harness its power and manifest your desires more quickly and easily.

I hope this step-by-step guide has been helpful to you. Remember, you are the creator of your own reality, and with the Law of Attraction, anything is possible. If you want more guidance on your manifestation journey, check out my other book, "369 Manifesting Journal: Your Manifestation Workbook Guided by the Law of Attraction."

Best wishes on your manifestation journey!

Printed in Great Britain
by Amazon